Problem-Based Learning

Engineering Lessons That Challenge

Table of Contents

Chapter 1. Introduction

Introducing our Special Report: "Problem-Based Learning: Engineering Lessons That Challenge!" Navigate through the fascinating world of engineering education as we delve deep into problem-based learning (PBL), a hands-on, student-centered approach that's transforming how we shape future engineers. This report doesn't just drone on about dull facts, but instead, it serves as your companion through lively case studies, interviews with experts, and a journey into classrooms where PBL is making a real difference. Crucially, it's written in a clear, everyday language that anyone can follow - yes, even if you can't tell a wrench from a ratchet! You don't have to be an Engineer to appreciate this report but you might just end up wishing you were one. Dive in, and let's embark on this extraordinary journey together!

Chapter 2. Unlocking the Mystery of Problem-Based Learning

Firstly, allow us to lay the groundwork. For a proper understanding of what problem-based learning (PBL) entails, you must firstly have a grasp of its concept and historical roots.

Problem-Based Learning (PBL), as a concept, was birthed around the late 1960s by Barrows and Tamblyn at the faculty of Medicine, McMaster University, Canada. Initially, this pedagogical approach was considered revolutionary, as it contrasted starkly with the traditional model of education; where students passively absorbed information delivered by an instructor. Akin to flipping a coin, PBL upended this traditional model, shifting the focus from teacher-led instruction to an active, student-centered learning process.

PBL immerses students in real-world problems, encouraging them to think, collaborate, and proactively participate in the quest for solutions. Unlike the conventional educational system, which often encourages an acquiesce-replicate approach, PBL spurs creativity, problem-solving skills, and the inherent potential for innovation within each student.

2.1. Understanding Problem-Based Learning

Now to the crux: How does PBL function in actual educational environments? A PBL classroom may seem confusing initially, especially to those accustomed to chalk-and-talk teaching. However, the apparent chaos serves as the breeding ground for active learning.

PBL classes seemingly follow a simple, yet profound process:

1. Students are presented with a complex, open-ended problem relevant to their subject.

2. The problem has no single solution, prompting students to research, propose, and iterate possible solutions.

3. Collaboration forms the bedrock of the process where students work in groups to solve the given problem.

4. The teacher serves as a facilitator, guiding discussions, refreshing knowledge, and providing feedback.

Think of the PBL approach as tour-guiding through the Grand Canyon. Instead of giving tourists a preset itinerary, the tour guide provides a map, basic tools for navigation, and lets the tourists explore their paths. When they stumble or struggle, the tour guide steps in to offer advice or lend a hand. By the end of the trip, tourists have learned live navigation skills, problem-solving, and teamwork; something a structured itinerary could never offer.

2.2. PBL in Engineering Education

Engineering is an epicenter of problem-solving. From creating advanced medical equipment to designing efficient cities, engineering wholly involves designing solutions. PBL, being steeped in problem-solving, fits into this landscape perfectly.

Engineering schools across the globe are embracing PBL. They recognize its power to turn classrooms into innovation labs. In these learning environments students experiment, fail, learn, and improve - mimicking the actual engineering workspace.

2.3. Unlocking PBL's Potential in Engineering

Unlocking PBL's real potential goes beyond introducing complex engineering problems to students. It requires careful creation of a conducive environment that fosters questioning, researching, failing, learning, and improving. It also calls for change in the role of the instructors, from lecturers to facilitators.

There are key ways to harness the true potential of PBL in engineering education:

1. Choose real-world, current engineering problems: Students' engagement levels rise significantly when faced with a problem that is both challenging and directly relatable.

2. Encourage critical thinking and creativity: Pushing students to apply theoretical knowledge in resolving real-life problems, drastically improves their ability to think critically and innovate.

3. Promote collaboration: The ability to function effectively in a group is a prized skill in today's workplaces. PBL inherently encourages collaboration.

4. Facilitators versus Instructors: Reinvention of the teachers' role is critical. Facilitators are required to encourage thinking and guide without spoon-feeding information.

2.4. Mastering PBL

Mastering PBL is not an overnight process; it requires time, effort, and patience. Both teachers and learners have to work in unison to ace this pedagogical model. Trained facilitators are required who can guide without controlling, foster a positive and encouraging classroom atmosphere, and give constructive feedback. Students, too, will undergo a metamorphosis; from passive learners to active contributors and problem solvers.

In the end, the true measure of PBL's distinction lies not in its uniqueness but rather in empowering each future engineer to actively shape the world. As PBL continues to revolutionize engineering education, the engineers of tomorrow are better equipped to address future challenges and create groundbreaking solutions. A learning style that opens up realms of possibilities, this is the true power of Problem-Based Learning.

Chapter 3. Theoretical Foundations of Problem-Based Learning

In this in-depth discussion on problem-based learning, we delve deep into the theoretical foundations that explain its efficacy. From the outset, it's important to note that problem-based learning (PBL) is not a wild shot in the dark, but instead, a method firmly built on the sturdy pillars of several learning theories. At the most basic level, these theories guide our understanding of how people learn and provide the necessary framework for implementing PBL in educational settings, especially in engineering.

3.1. The Origins of PBL

PBL originated as a medical education approach at McGill University in the 1960's, aimed to better prepare students for the complex problem solving required in clinical medicine. The need to decrypt and analyze real-life problems are just as crucial in engineering, fanning the adoption of PBL across disciplines including engineering education. To fully comprehend PBL, it's integral to understand the key theories that underpin it: constructivism and situated cognition.

3.2. Constructivism and PBL

Constructivism is a learning theory based on the assumption that humans generate knowledge and meaning from their experiences. Piaget and Vygotsky, two influential theorists of constructivism, both argued that knowledge is actively constructed in the mind of the learner, and not passively received from an external source.

In context of PBL, the problems students encounter are open-ended, mirroring the complexity of real-world problems. This prompts them to pull together their existing knowledge, and acquire newer information to tackle the problem, facilitating 'constructive' learning processes. PBL aligns with the constructivist creed ". . . that learning is an active, constructive process" (Bruner, 1986), by leveraging real-world problems in the classroom, to ignite student autonomy and personal growth.

3.3. Situated Cognition and PBL

Coined by Brown, Collins, and Duguid in their 1989 seminal paper, situated cognition argues that knowledge is bound to the context of its learning and use. They maintain that authentic activities, or tasks performed by practitioners in the field, greatly aid learning.

In PBL, the 'problems' posed to learners are often based on real-life scenarios, forcing students to think and act like practitioners. It aligns with Lave and Wenger's theory of "legitimate peripheral participation" wherein learners gradually immerse themselves in a community of practice, moving from being novices to experts. By implementing PBL, classrooms become labs where theories meet practice, simulating the cognitive experiences of working engineers.

3.4. The Role of Metacognition in PBL

'Reflection' and 'self-regulation' are two components of metacognition, the understanding and control of one's own learning, which are integral to PBL. Reflection enables learners to recognize the gaps in their understanding and redirect their learning processes efficiently. Self-regulation, on the other hand, involves planning, monitoring, and assessing one's learning.

By this merit, PBL cultivates metacognitive skills as it requires students to regulate their progress, given the self-directed nature of PBL. The cycle of hypothesizing, receiving feedback, and making necessary adjustments to solve complex problems feeds into metacognitive processes, resulting in refined problem-solving abilities and enhanced understanding.

3.5. Collaborative Learning in PBL

Collaborative learning theory, with its grounding in Vygotsky's constructivist approach, lends to the collaborative aspect of PBL. It surmises that learning is socially mediated and best achieved through others' help. In PBL, students often work in teams, discussing ideas, challenging each other, and making collective decisions, which brings a collective intelligence to solving complex problems.

This social interaction not only promotes comprehensive understanding but also encourages students to become more responsible learners. Through group work, students learn from each other and develop transferable skills such as communication, teamwork, and conflict resolution, which are vital in real-world engineering.

To wrap, the theoretical foundations of PBL amalgamate different but complementary learning theories. Constructivism and situated cognition infuse authenticity into the learning process while collaborative learning and metacognitive theories amplify the self-directed and social aspects of learning. Together, they offer a powerful and rigorous framework that makes PBL an effective tool in engineering education and beyond. By implementing PBL, classrooms truly become fertile grounds where future engineers are nurtured and readied to face real-world challenges.

With this foundation in place, we are better equipped to understand the hows and whys of PBL implementation in the engineering classroom, as well as its impact on students. This understanding can also serve as a guide in moulding custom strategies to optimize the delivery and outcome of PBL based courses for future engineers.

Chapter 4. Engineering Education: The Traditional Methods vs. PBL

The foundation of engineering education has traditionally rested on firm pillars: mathematics, physics, and a smattering of domain-specific subjects, all delivered through an instructor-led, lecture-centric approach. This method, apt as it may seem, has its own set of limitations, often resulting in engineers who may have mastered theoretical concepts but falter when it comes to practical, real-world problems. Enter Problem-Based Learning (PBL), a dynamic, student-centered learning philosophy that is turning this traditional model on its head.

4.1. The Traditional Approach

In the conventional model of engineering education, classes are inherently teacher-centric. An educator takes to the dais, armed with a well-crafted lesson plan centring on a specific topic. Through carefully chosen examples and explanations, the teacher dispenses knowledge, while the students receive it passively. The focus is primarily on the absorption of factual knowledge and procedural steps.

The backbone of this passive learning style is rote memorization, aimed at storing data to be vomited back up during examinations. Here, students can succeed by simply remembering the correct answers to specific questions. Little attention is paid to fostering critical thinking, creativity, and the ability to apply knowledge to novel situations.

4.2. Practical Implications and Shortcomings

The traditional method puts a wall between academic learning and its practical implementation. It fails to illustrate the real-life concepts that engineers will eventually have to grapple with. These are the environments where textbook problems — those that have a single correct answer, obtainable by a well-defined series of steps — are a rarity.

Traditional learning approaches can also foster an environment of isolation, as students often end up working individually. It ignores a crucial aspect of real-world engineering — team collaboration. Real-life engineering projects require engineers to work as teams, integrating ideas and skills from multiple domains.

Furthermore, when learning is time-bound by the structure of term-end exams, deep understanding can get sidetracked for rote memorization under the pressure of grades. So while students may academically excel, they may still be unequipped to deal with real-world challenges beyond their textbooks.

4.3. Introduction to Problem-Based Learning (PBL)

PBL offers a contrasting model to the traditional approach. This method introduces students to real-world engineering problems from the get-go. Students are split into teams and tasked with a specific problem that needs a creative solution. This problem forms the basis for their learning.

What sets this approach apart is that it flips the classroom model: the instructors act as facilitators or coaches, providing guidance when necessary, while real learning happens proactively by the students.

They dive into research, brainstorm, develop hypotheses, test their solutions, and present their findings — all by themselves.

It promotes self-directed learning, enhances team-working and communication skills, and develops the ability to apply knowledge to unfamiliar situations — the real essence of engineering.

4.4. Benefits of PBL

The biggest advantage of PBL is its focus on active learning. This promises a more engaged class — students are no longer passive recipients of knowledge. They are researchers, practitioners, and presenters. This not only heightens their interest but also enhances their understanding of the subject matter, ultimately leading to better retention.

Working in groups, students understand the power of teamwork and collaboration. They acquire critical soft skills, such as effective communication, collaboration, creativity, and in turn, they become better prepared for the real engineering world out there.

Moreover, PBL aligns closely with real-life engineering tasks. Students can make connections between their course content and engineering practice. They understand the pragmatic aspects of what they learn and hence tend to be better engineers in practice.

4.5. PBL vs Traditional: Finding a Balance

Despite the apparent advantages, PBL may not always entirely replace traditional education methods. Abstract concepts often require more explicit instructional strategies than PBL provides. Here, the traditional approach can deliver foundational understanding effectively.

Ideally, a mixture of both models suits best — traditional education to establish foundational knowledge and PBL to build on this foundation through problem-solving, collaboration and real-world applications.

PBL and traditional engineering education can be seen not as competing alternatives but as complementary parts of a whole. Together, they form a complete model of engineering instruction — one that equips students with the skills necessary to become globally competitive engineers.

In the fast-evolving world of technology, the ability to problem-solve and innovate is more crucial than ever. PBL is a potent tool in this regard. By effectively adopting and adapting both methods, we can train engineers who are not only sound in theoretical knowledge but also in its practical application. After all, engineering is not just about learning principles but also their prudent application.

Chapter 5. The Implementation Process: Embedding PBL in Engineering Courses

5.1. Understanding PBL

Let's start from the ground up and first understand what PBL is before plunging into how it finds a fit in engineering education. Problem-Based Learning (PBL) is a pedagogical method where students learn about a subject through the experience of problem-solving.

Simply put, students gain new knowledge and skills by tackling real-world challenges and they learn not only from the teacher but from the process of resolving a problem and from their peers as well. PBL promotes an active learning environment where the students have ownership of their learning, making them more personally invested in the process.

5.2. Transitioning from Traditional Teaching to PBL

Transitioning to PBL in engineering courses is not a simple switch. It requires a systematic change in both curriculum design and teaching methods. There are several steps and processes ingrained in this transformation: faculty development, course redesign, identification of real-world problems, etc.

100% institutional buy-in is necessary for PBL implementation to be successful. Everyone involved, from administrators to faculty, must understand and support the approach to facilitate PBL implementation smoothly.

A PBL curriculum differs significantly from a traditional lecture format. Instead of conveying knowledge primarily through lectures, seeds of knowledge are sown in students through problems directly aligned with real-world situations. This flips the classroom model from a teacher-centered to a learner-centered model, where students are encouraged to explore topics independently and collaboratively.

5.3. Faculty Development

A significant enabler for the successful institution of PBL is developing faculty capacities. Educators must transition from the traditional role of instructors to that of facilitators in a problem-based setting. Alongside their subject knowledge, they must also be adept at fostering critical thinking, collaboration, and research skills in students.

Regular faculty workshops can provide the essential skill sets needed in the PBL realm. Including teaching teamwork, questioning techniques, designing real-world challenges, identifying learning objectives, and evaluating student performance in a PBL scope. For this purpose, collaborations with more experienced PBL institutions can prove beneficial in guiding the transition.

5.4. Course Redesign

With the adoption of PBL, the engineering curriculum will need a considerable overhaul. Course redesign will involve mapping out learning goals and objectives, selecting and creating problem scenarios, and integrating evaluation methods.

The learning goals need to be student-centered and should encourage action rather than recall. Next comes the problem selection and development. The problems chosen should align with the course objectives and be complex enough to be engaging. They should also be based in real-world scenarios to give students exposure to actual engineering situations.

The assessment framework must also shift to accommodate new learning goals, focusing on a student's ability to apply, analyze and evaluate knowledge rather than just recall facts.

5.5. Integration of Real-World Problems and Case Studies

To gain truly meaningful insights, students should engage with problems that bear the weight of reality. PBL encourages students to tackle real-world challenges, ones that engineers face daily. These problems should be open-ended, have more than one feasible solution, and involve constraints and trade-offs. They should mimic the complexity and ambiguity of real engineering problems.

Curriculum designers can use case studies from industries, relevant research, global engineering issues, failed projects, etc. to develop these problems. A crucial aspect is to design problems that encourage learners to seek out and learn concepts they need to solve them, thereby enhancing their information processing skills.

5.6. Assessment and Feedback in PBL

Evaluation is an essential part of any teaching methodology. In PBL, it shifts from conventional testing methods to more holistic ones. Assessment must evaluate the depth and breadth of knowledge acquired, problem-solving strategies used, and the degree of collaboration among students.

Assessments in PBL mirror the teaching approach. They, too, should be carefully designed to reflect real-world situations. Components of assessment include: * Individual portfolios * Team projects * Oral presentations * Reflective statements

The assessments, carefully blended in with a traditional grading approach, can provide a multi-dimensional measure of student achievement.

Lastly, feedback plays a vital role in PBL. Students should be encouraged to give and receive feedback among their peers, facilitating continual improvement in problem-solving skills and enhancing a collaborative learning environment.

5.7. Concluding Remarks

As with any significant educational shift, embedding PBL into engineering courses may never be an entirely smooth transition. Challenges will arise, whether they be logistical, conceptual, or practical. But through a clear understanding of the purpose and methodology of PBL, these hurdles can be overcome, leading to a valuable and profound impact on teaching and learning alike.

The adaptation of PBL in engineering is not just a change in method, but a paradigm shift. With these guidelines, the path to successful implementation and the realisation of this shift is made more attainable. Thus, equipping future engineers with skills that go beyond technical abilities, ready to solve real-world problems and contribute positively to society.

Chapter 6. Case Studies: Success Stories in Problem-Based Learning

Combining rigorous academics with relevant, hands-on application, problem-based learning (PBL) is a pedagogical method that's creating waves across engineering education worldwide. Let's delve into some success stories that exemplify the transformative power of this approach.

6.1. Recreating a Bustling City: Civil Engineering PBL at MIT

At MIT, one notable example of civil engineering problem-based learning spanned an entire semester, as students were tasked with designing a full-scale city layout from scratch. Based on realistic constraints such as budget restrictions, zoning regulations, and environmental implications, students worked in teams to brainstorm solutions, analyze potential issues, and draft comprehensive city plans. Such immersive, hands-on lessons promoted not just theoretical understanding but also practical insight into the complex realities of civil engineering, preparing the students for real-world challenges.

6.2. Future-Proof Farms: A Lesson in Agricultural Engineering at UC Davis

UC Davis's College of Engineering implemented PBL to teach students about sustainable agricultural practices. In this course, students were required to design a viable and environmentally-friendly farming system by considering various constraints like soil type, weather patterns, and crop rotation schedules. This interactive learning experience promoted innovation and critical thinking, empowering students to create projects that had the potential for real-life implementation. The course evolved into a platform fostering young entrepreneurs with innovative farming solutions, many of which have been granted patents.

6.3. Robotic Explorations: The Rise of Mechatronics at Stanford

Stanford University's School of Engineering offered a mechatronics course where PBL took center stage. Here, students were given the challenging task of creating a fully functioning robot. The process involved designing, building, and testing their robot, with particular focus on tasks like overcoming obstructions and autonomous navigation. This learner-centered approach allowed students to directly apply theories to practical situations, enhancing their problem-solving skills while providing real-world context to theoretical learning.

6.4. Caring for the Earth: A Environmental Engineering PBL at Yale

The Department of Chemical and Environmental Engineering at Yale University took to PBL to teach students about reducing and managing waste production. The challenge presented to students involved designing a waste management plan for a mid-sized city with increasing population growth and limited landfill space. Through this PBL course, students tackled pressing environmental concerns while gaining a deeper understanding of the social, economic, and environmental complexities underpinning such tasks.

=== Faultless Mechanisms: Mechanical Engineering PBL at Imperial College London

Imperial College London's Department of Mechanical Engineering introduced a PBL program where students banded together to solve a specific problem. Given the task of designing a safety mechanism for an industrial machine, students researched existing safety protocols, pondered potential mechanical failures, and took a hands-on approach to devise a solution. This problem-oriented approach to learning resulted in students producing innovative designs that were applicable to real industrial problems.

6.5. Digital Landscapes: Computer Engineering PBL at Singapore's NUS

The National University of Singapore (NUS) embraced the PBL methodology within its computer engineering program, where students undertook a project to develop gaming software. By undertaking the task from system design to execution, students experienced the rigorous process of software development. Not only did this provide an applied learning scenario to understand coding theories, but it also nurtured creativity and teamwork, preparing students for the future tech industry.

These examples demonstrate how PBL has helped students gain a deeper understanding of engineering concepts and their practical applications, thus bridging the gap between academic theories and real-world applications. With PBL, students do not just memorize information; they immerse themselves in complex problems that require critical thinking, creativity, and a solid understanding of engineering principles. And in such pioneering institutions, students graduate not just with degrees but with experiences and real-world solutions that can impact their chosen fields and the world at large.

Chapter 7. Interviews: Educators Advancing PBL in the Engineering Classroom

Educators play an undeniably pivotal role in the deployment of Problem-Based Learning (PBL) within the sphere of engineering education. Their influence, experiences, and innovations are worthy of exploration and discussion. Let's dissect the various perspectives they bring to the table and how they're making PBL a permanent fixture in engineering education.

7.1. Defining Problem-Based Learning (PBL)

But first, what exactly is PBL? It's a teaching method that presents students with complex, real-world problems as a way to stimulate critical thinking, problem-solving skills, and a deeper understanding of the theoretical concepts. In PBL, beginning with a problem is essentially the 'problem' - an exciting shift from traditional teaching methodologies. But how do educators integrate this approach into their classrooms?

7.2. Adoption and Integration of PBL

Maria Thompson, a veteran mechanical engineering professor from Northern Tech University, began incorporating PBL into her curriculum just five years ago. Today, her courses are unique amalgamations of theoretical instruction and practical, problem-based application. Her PBL adoption story serves as a roadmap for many educators wishing to make a similar transition.

"Shifting to PBL wasn't easy," Thompson begins. "It required a paradigm shift - focusing on principles first, then diving into practical case studies. My first PBL trial was integrating a local factory's equipment optimization challenges into our coursework."

As Thompson explains, her biggest hurdle was not introducing the real-world problems into the coursework, but rather getting the students to actively engage with the problems. "PBL demands active learner engagement, far removed from the rote learning many students are accustomed to," she explains.

7.3. Variants of PBL in Practice

Different educators adopt PBL differently. Let's explore two other variants of PBL implementation.

Dr. Antonio Ramírez, an electrical engineering professor at The Polytechnic School in Madrid, introduced a variation of PBL called Project-Based Learning. In this offshoot method, students work on a project over an extended period that often involves solving a complex, multidisciplinary problem. Ramírez's senior students, for instance, work on a semester-long project involving the design and implementation of an energy-efficient home automation system.

Conversely, Dr. Elise Kim from Korea's Seoul Tech integrates PBL through an integrated approach. She places emphasis on equipping students with basic knowledge and then gradually introduces complex practical problems. Thus, for her, PBL is an iterative process of problem identification, concept learning, and application – a cycle that she feels gives students a comprehensive understanding of engineering principles.

7.4. Overcoming the Challenges

PBL often involves a learning curve not just for students but also for educators. Dr. Kim noted, "Students initially resisted this restructured way of teaching. However, after seeing improvements on tests and realizing they were more equipped for real world challenges, they were more open to the change."

Adapting to the PBL model requires substantial effort from educators, as they must constantly update their problem sets to reflect evolving real-world challenges. Thus, the upkeep and selection of problems require significant time and effort. Nonetheless, educators are generally optimistic and believe that the extra effort is a worthy investment in shaping future engineers.

7.5. Technologies For PBL Implementation

Integrating technology into PBL is another emerging trend. Dr. Ryan Peterson from California's Tech Coast University uses sophisticated simulation software to bring real world problems directly into his classroom. This has allowed his students to design, simulate, and evaluate complex engineering systems - all within a controlled, educational setting.

Maria Thompson noted that digital communication platforms, particularly during the pandemic, have become a boon for PBL. "Students collaborate online to solve problems, altering how we view 'classroom' settings. Additionally, digital resources readily complement PBL, giving students access to a wealth of knowledge beyond textbooks."

7.6. The Future of PBL in Engineering Education

Despite initial resistance and challenges, PBL is fast becoming a definitive part of engineering education. "The integration is gradual,"

Thompson says, "but I foresee PBL becoming an integral part of the engineering curriculum in the next decade." With the combined efforts of educators like Thompson, Kim, Ramírez, and Peterson, this prediction seems more plausible each day.

As we progress, PBL's prominence in engineering education seems destined to rise. This shift is an essential step in preparing future engineers to face real-world challenges, arming them not just with theoretical knowledge but also with the crucial ability to find practical solutions to complex problems. And as our educators have shared, guiding the engineers of the future makes every challenge worth it.

Chapter 8. Challenges & Solutions in Problem-Based Learning

Problem-based learning (PBL) represents a paradigm shift in educational methodology. By focusing on hands-on, practical experiences, PBL seeks to equip students with the tools they need to tackle problems creatively and effectively. However, as with any significant change in educational approach, PBL also presents a unique set of challenges.

8.1. The Challenge of Student Engagement

Student engagement goes beyond simply paying attention in class. A truly engaged student is invested in their learning - they take ownership, are motivated to learn, and view their education as meaningful. For students accustomed to traditional, lecture-based formats, PBL can initially seem overwhelming.

Firstly, the demands for self-directed study can cause anxiety in students who are more comfortable with clear directives and structured learning environments. Additionally, the focus on group work can lead to problems such as freeloading, where some group members contribute less than others. Finally, weaker students may feel neglected as teachers give more attention to students who are active and confident.

But there are solutions to these challenges.

8.2. Activating Intrinsic Motivation

To alleviate anxiety and foster engagement, educators need to activate students' intrinsic motivation. This refers to motivation that comes from within the student themselves, as opposed to extrinsic motivation which is driven by external factors, like grades or approval.

Active learning strategies such as questioning, group discussions, brainstorming, and peer teaching can all contribute to creating an engaging PBL environment. Providing guided autonomy, where educators support students in managing their own learning while also providing structure, can both alleviate anxiety and foster engagement.

8.3. Encouraging Balanced Contribution in Group Work

In terms of group work, educators need to proactively manage group dynamics to ensure equal contribution from all members. This could involve assigning roles within the group, monitoring group progress regularly, and assessing individual and group performance.

8.4. Supportive Environment for Weaker Students

For weaker students, special provisions can be made. Reinforcing basic concepts, organizing remedial classes, and using scaffolding techniques can all make a significant difference. The goal should be to provide a supportive learning environment that also challenges every student at their level.

8.5. Resistance from Educators

Despite the proven benefits of PBL, there continues to be resistance from educators who perceive it to be less effective than traditional methods, too time-consuming, or hard to assess.

The shift to a student-centered approach where teachers become facilitators rather than sole providers of information requires significant adjustment. Additionally, planning and implementing PBL can be time-consuming and requires resources that are not always readily available.

8.6. Professional Development and Support

To overcome these challenges, institutions need to invest in professional development for educators. This could be in the form of training programs, peer-support networks, and coaching from experienced PBL practitioners. Administrations should ensure they provide the necessary resources, both in terms of materials and time.

8.7. Developing Assessment Methods

Assessment in PBL poses another challenge, as learning goes beyond memorization of facts, with a focus on skills and application. Traditional exams might not work, requiring innovative alternative assessment methods, like reflective journals, rubrics, and peer-assessments.

8.8. Institutional Adaptability

Lastly, institutions themselves need to adapt. Traditional systems and structures may no longer fit with a PBL approach. For instance, rigid timetables may need to be revised to accommodate longer, more fluid PBL sessions.

In conclusion, while problem-based learning presents significant challenges, these can be overcome with effective strategies and a commitment to adopting learner-centered approaches. Through harnessing the full potential of PBL, we can transform engineering education, equipping the next generation of engineers with the skills they need to innovate and succeed.

Chapter 9. Assessing the Impact: How Effective is PBL?

The exploration of the effectiveness of problem-based learning (PBL) in engineering education is predicated on several fronts. Empirical research, observation, and anecdotal evidence spread over multiple disciplines have come together to form a comprehensive understanding of PBL's impact.

9.1. Quantifying PBL Impact through Research

Research-centered assessment is one of the objective ways to unwrap the benefits and shortcomings of PBL in engineering education. A variety of studies across different educational institutions and geographies have provided insights critical to assessing PBL's effectiveness.

One such study, conducted by Dochy, et al., employed a meta-analysis of 43 empirical studies on problem-based learning. Their findings revealed that PBL students tend to perform better in skills-based assessments than their traditional instructional methods counterparts. However, results were mixed in terms of knowledge retention, where PBL students performed slightly lower than their peers taught via the lecture-based method.

Another important study by Strobel and van Barneveld consolidated findings from previous research to understand PBL's long-term benefits. They discovered that PBL students tend to outperform non-PBL students in professional skills such as collaboration, self-directed learning, and problem-solving.

9.2. Real-World Application: Evidence from the Field

While research provides invaluable data-driven insights, real-world observations and anecdotal evidence provide an on-ground perspective often absent in studies conducted in controlled environments. By observing PBL in the application, we can gain a granular understanding of how it shapes the learning process.

Educators and researchers at Purdue University School of Engineering Education have noted that PBL has given students a significant edge in job interviews and the workforce due to their extensive problem-solving experience and collaboration skills.

Moreover, anecdotal evidence from institutions using PBL suggests improvement in students' enthusiasm and engagement. Teachers report that students in PBL classrooms are often more eager to participate and take control of their learning.

9.3. Comparing PBL with Traditional Learning Approaches

Comparing PBL with other instructional methods is another way to gauge its effectiveness. By juxtaposing the outcomes, advantages, and shortcomings of PBL alongside traditional formats, we can determine where PBL shines and where it may need augmentation.

One of the key areas where PBL outperforms traditional learning is in the application of theoretical knowledge to real-world scenarios. This practical approach provides students with better preparation for their future professional roles.

However, in the context of content-heavy and foundational subjects, traditional learning approaches can ensure micro-learning outcomes better. In contrast, PBL might need a solid basis of initial knowledge to build upon with its problem-solving focus.

9.4. Surveying Students and Teachers: The Perception of PBL

Lastly, assessing the effectiveness of PBL would be incomplete without considering the perceptions of students and educators themselves.

Surveys of students reveal mixed reactions to PBL. Some students, especially those fond of active and self-directed learning, express great satisfaction and the discovery of enhanced learning outcomes. However, some students cite feelings of initial discomfort and struggle due to the shift from traditional instructor-centric learning.

For educators, embracing PBL often necessitates a significant shift in teaching methodology. Some teachers have noted the need for extensive training and preparation to implement PBL effectively. Yet, many agree that the benefits of this approach, such as increased student engagement and superior learning outcomes, make the transition worth it.

In conclusion, the effectiveness of PBL in engineering education is multi-faceted. While empirical research provides comprehensive insights, anecdotal evidence, field observations, and personal testimonies are equally important. By continuosly assessing and refining PBL, we can make strides towards transforming engineering education.

Chapter 10. Future Prospects: The Evolution of Problem-Based Learning

In the swiftly evolving landscape of 21st-century education, problem-based learning (PBL) has become an indispensable paradigm. This innovative learning approach, which merges real-world problems with the rigors of intellectual inquiry, has revitalized engineering education, creating a new breed of engineers equipped to tackle the multifaceted challenges of the age.

10.1. The Essence of Problem-Based Learning

Problem-Based Learning, at its core, is an instructional methodology that encourages learning by engaging students in real-world problems without predetermined solutions. This pedagogical approach fosters pivotal skills necessary for engineers such as critical thinking, problem-solving, creativity, and teamwork.

The immersion in lifelike scenarios helps students to acquire and apply fundamental knowledge in engineering, thus giving them a more significant understanding of how theories are applied in practical, real-life situations.

10.2. Historical Traces: Roots and Development of PBL

The cradle of PBL is found in McMaster University's medical program in the late 1960s. The frustrations of the faculty regarding traditional pedagogy sparked an education revolution that would later impact not only medical training but also various disciplines, including engineering.

Throughout the 1970s and 1980s, PBL principles were gradually implemented in different courses worldwide, with Aalborg University in Denmark being a noted trailblazer within engineering education. They recognized early on how perfectly PBL aligns with the quintessence of engineering: solving complex problems.

10.3. Embracing Change: Trends and Challenges in PBL Implementation

Nonetheless, while a historical perspective provides context, the future of PBL depends on how contemporary educational institutions adapt it to the demanding, ever-changing landscapes of technology and society.

Despite the numerous well-documented benefits of PBL, its implementation has not been without challenges. Critics say it's time-consuming, requires significant resources, and can be challenging to assess. Institutions also grapple with resistance from educators entrenched in traditional pedagogical methods who find PBL intimidating or impractical.

Despite these hurdles, the rise of digital technologies and e-learning platforms provides new avenues for the implementation of PBL, making it increasingly possible to offer this immersive, collaborative learning model on a substantial scale.

10.4. Projections: Future Trajectories in PBL

The future of PBL in engineering education is promising. By intertwining technological advancements with innovative pedagogical practices, PBL is set to further extend its transformative impact.

The rise of artificial intelligence, in particular, promises to revolutionize PBL. Machine learning algorithms may soon customize real-world problem scenarios to the unique learning styles and speeds of individual students. Simultaneously, AI will streamline the assessment process, evaluating student performance in a way that's more nuanced and responsive than traditional grading methodologies.

10.5. PBL and the Industry-Academy Bridge

Arguably, the most enticing advantage of PBL is its potential to reduce the disconnect between industry and academia. Industrial professionals consistently lament that newly recruited engineering graduates lack the practical skills necessary for solving real-world problems. PBL mitigates this by fostering such skills throughout the educational process, ensuring that students graduate as problem solvers ready for industry challenges.

10.6. Towards a PBL-centric Future

In conclusion, the future prospects of Problem-Based Learning in engineering education are vibrant. PBL is not a trend that will fade; instead, it has demonstrated its value as a powerful tool equipping engineering students with the necessary skills to face complex, real-world scenarios.

The journey towards a PBL-centric curriculum will not be without its ups and downs. Nonetheless, the compelling advantages of PBL - the cultivation of problem-solving skills, the bridging of the academic-industry gap, the affinity with technological advancements - ensures it is a voyage worth embarking on. By embracing PBL, the future of engineering education is bound to be a thrilling one that consistently edges closer to the realities of the profession.

35

Chapter 11. Next Steps: Incorporating PBL into Your Engineering Curriculum

As we step into a future where complex and diverse challenges await, it is essential for the next generation of engineers to be fully equipped to face these problems confidently. The integration of Problem Based Learning (PBL) into the engineering curriculum can make this vision a reality. The exhaustive content below discusses the "next steps" to consider to successfully incorporate PBL in your engineering curriculum.

11.1. Understanding the Basics of PBL

Before the integration, the first step is to have a clear understanding of PBL, its principles, and its objectives. PBL finds its roots in Constructivist Pedagogy emphasizing that learners construct knowledge actively instead of merely receiving it passively. It's a dynamic form of learning where students involve themselves in real-world problems and through reflection and problem-solving, relates the knowledge to the context.

11.2. Evaluating Current Curriculum Models

Carrying out a scrutinizing evaluation of the current curriculum models in place is crucial. This evaluation will help in identifying the gaps where PBL can assist. It's also essential to analyze how compatible the existing curriculum is with PBL; some curricular may require more extensive changes than others.

11.3. Building Capacity: Training Educators

Educators play a crucial role in PBL. They're not just deliverers of knowledge but facilitators of learning. Thus, adequate training must be provided to teachers to ensure they can support the implementation of PBL effectively. This also includes orienting them towards a more open-ended and facilitative style of teaching rather than a directed one.

11.4. Designing PBL-Compatible Problems

A successful translation of PBL into practice largely depends on the design of problems or projects. These problems should be complex enough to challenge students yet manageable within the realistic constraints of the classroom. They should be sufficiently ill-structured, mirroring real-world problems, and should cut across the boundaries of traditional subjects or disciplines.

11.5. Developing Assessment Strategies

Unique modes of instruction, such as PBL, require a different approach to assessment. The typical paper-and-pencil examinations may not do justice to assessing the outcomes. Hence, developing qualitative assessment strategies that not just evaluate a student's ability to memorize data and facts but also their problem-solving capabilities, collaborative skills, and ability to apply knowledge is crucial.

Learning Objectives	Assessment Method	Student Product	Feedback
Understanding of key concepts	Concept Maps	A network of interconnected concepts	Detailed, supportive critique
Application of knowledge to new situations	Working Prototype	A working model answering the problem	Observations, suggestions for improvements
Collaboration	Peer Evaluation	Individual and group reflection	Constructive feedback, self-improvement goals
Interdisciplinary thinking	Research Presentation	Presentation showcasing interdisciplinary solutions	Questions, elaboration on key ideas

11.6. Fostering a Supportive Learning Environment

For PBL to be functional and effective, it is crucial to cultivate a conducive learning environment where students enjoy collaborative problem-solving and take responsibility for their learning. Flexibility in teaching methods and a subtle shift from a teacher-centered to a student-centered learning environment can foster this change.

11.7. Continuous Evaluation and Revision

It is fundamental to continuously monitor the effectiveness of PBL integration into the curriculum. This monitoring can be done through student surveys, teacher feedback, and evaluation of student performance. Based on these evaluations, necessary revisions should be made for further improvement.

By following these comprehensive steps, the integration of PBL into your engineering curriculum can come to fruition. The result? A new generation of engineers eager to solve real-world problems in their respective fields.

It's not a small task; indeed, integrating PBL needs effort and time. Still, for the rewarding future it promises - a stream of mindful and skilled engineers solving world problems - it's an expedition worth embarking on.